Andrea Faussone

CYLINDRICAL COMPRESSION HELIX SPRINGS FOR SUSPENSION SYSTEMS

design requirements and calculation nomogram

Copyright © 2018 Andrea Faussone

All rights reserved.

Cover Copyright © 2018 Andrea Faussone

All rights reserved.

UUID: c3f5307a-b1af-11e8-881a-17532927e555

This ebook was created with StreetLib Write
http://write.streetlib.com

Table of contents

1.0 PURPOSE	1
2.0 REFERENCES	2
3.0 APPLICABILITY	3
4.0 EXAMPLE OF DESIGN PRESCRIPTIONS	4
4.1 Symbols	4
4.2 Generalities, definitions and tolerances	4
4.3 example of spring suspension designation	10
5.0 CALCULATION NOMOGRAM	12
5.1 Purpose	12
5.2 Applicability	12
5.3 Characteristics identified by the nomogram	12
5.4 Calculation Example	14

1.0 PURPOSE

The purpose of this publication is giving to the designer an example of the characteristics normally reported on the technical drawings of the cylindrical compression springs for suspension systems and an example of a calculation nomogram.

The usual tolerance fields for correct production are also provided. The men who wish to deepen the calculation and design methods, refer to the following point.

2.0 REFERENCES

This publication agrees with the following design and calculation rules:

- UNI – 7900 Part 2
- **Helical compression cylindrical springs - design, calculation and verification by Andrea Faussone, Engineer (2018)**
- Metric Decimal System

3.0 APPLICABILITY

This standard relates to hot-rolled cylindrical compression springs, used for the suspension systems of road vehicles, off-road vehicles and vehicles on rails.

4.0 EXAMPLE OF DESIGN PRESCRIPTIONS

4.1 Symbols

D = average diameter of the spring under load, in mm
D_e = outer diameter of the spring, in mm
D_i = inner diameter of the spring, in mm
d = nominal diameter of the wire, in mm
L_a = contact surface length, in mm
L_o = free length of the spring, in mm
L_1 = spring length at static load, in mm
L_2 = spring length at dynamic load, in mm
L_b = block length of the spring, in mm
P_1 = static load, in N
P_2 = dynamic load, in N
P_c = test load, in N
$C = \Delta f/\Delta P$ = spring flexibility, in mm/N
$\Delta f = (L_1 + 25) - (L_1 - 25)$ = spring elastic displacement, in mm, corresponding to the difference of loads ΔP
ΔP = load variation, in N, measured on lengths $L_1 - 25$ e $L_1 + 25$
ΔP_1 = tolerance on the static load P_1, in N
I = number of active coils
I_t = total number of coils
a = inclination of the end coil, in mm
b = thickness of the coil at the tapered ends, in mm
p = coefficient for the calculation of the tolerance on the form error
s = tangential projection of the end of the spring, in mm
α = angle relative to the mutual position of the ends of the coils
$\Delta\alpha$ = tolerance on the α angle

4.2 Generalities, definitions and tolerances

4.2.2 Round bar
As a rule, the cylindrical compression springs for suspension systems of motor vehicles are obtained from ground round bars.

4.2.3 Diameter of the spring
For cylindrical compression springs for suspension systems, only the internal diameter D_i must be specified.

4.2.4 Ends types
If necessary, the supporting surfaces of the tapered end springs (type 3 and 4) can be ground.

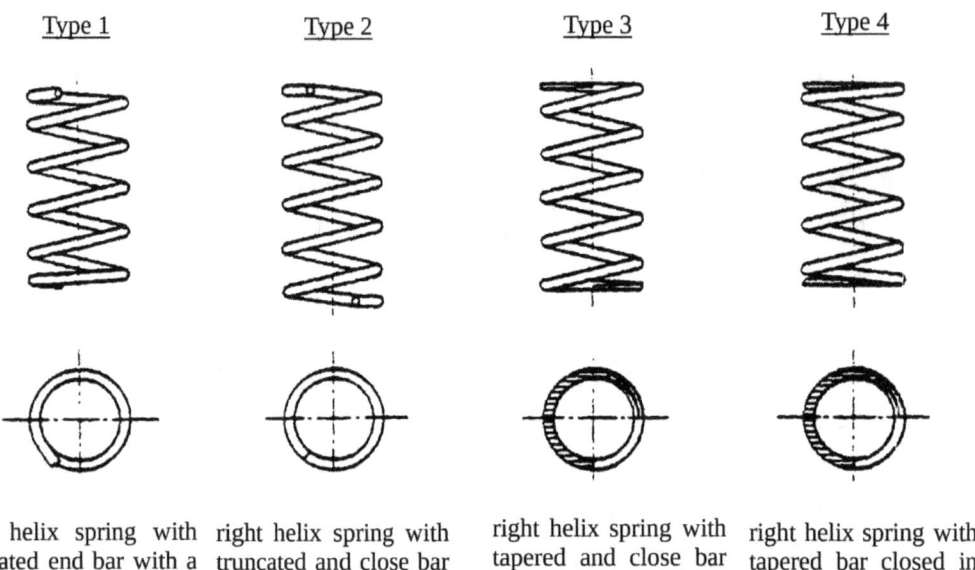

right helix spring with truncated end bar with a tangential and close projection

right helix spring with truncated and close bar end

right helix spring with tapered and close bar end

right helix spring with tapered bar closed in contact

Figure 1-Ends Type

4.2.5 Total coils "I_t"

It is the total number of complete coils and coil fractions. To facilitate the manufacture of the springs it is recommended that the coil fractions are limited to 1/4, 1/2 and 3/4, with preference to 1/4 and 3/4 of coils.

4.2.6 Active coils "I"

It is the total number of coils less the number of coils at the spring ends that remain inactive. As a rule, the ends and seats of the springs are constructed in the manner to have from 2/3 to 1 inactive coil at each end of the spring.

The number of active spring coils, with various types of ends, must be determined as follows:

- Springs with type 1 and 2 ends: the number of active coils is in relation to the portion of the end coils that are in contact with the lower and upper support surfaces. For this reason, the number of active coils is defined case by case based on the design of the spring.
- Spring with ends type 3: $I \approx I_t - 1,5$
- Spring with ends type 4: $I \approx I_t - 2$

4.2.7 Free length "L_0"

It's the length of the unloaded and settled spring, measured parallel to the axis of the spring in correspondence with the extreme points or the support surfaces. The free length must be, except for special needs, prescribed in the drawing with the sign ≈ (that means "approximately").

4.2.8 Static load length "L_1"

It's the length of the spring under the load corresponding to the weight of the vehicle stopped at full load. For light vehicles systems, it's the spring length under the load corresponding to the standard weight.

4.2.9 Dynamic load length "L_2"

It's the minimum length that the working spring can take. A sufficient light must be provided between the individual coils to ensure that no active coil comes into contact due to errors in the helix pitch.

4.2.10 Block length "L_b"

It's the length of the spring with all the coils in contact, measured parallel to the axis of the spring in correspondence with the support points or support surfaces. Normally, to carry out this measurement, it is necessary to apply to the spring a load equal to 150/100 of the one above which there is no further yielding. The block length is approximately obtained from the following formulas:

- Springs with end types 1 and 2 $L_b \approx (i_t + 1) \cdot d$
- Springs with end types 3 and 4 $L_b \approx (i_t - 0,5) \cdot d$

The block length must be indicated on the drawing with the sign ≈ (approx.). If verification is necessary, the block length must be prescribed as a maximum value in the drawing. In this case, when calculating the maximum block length, the tolerance on the wire diameter, the possible coating and the angular tolerance on the mutual position of the ends of the spring must be taken into account.

4.2.11 Test load "P_c"

It's the load to which the spring must be stressed 3 times before checking the dimensions and loads. For the suspension springs of the cars, it is equal to the load calculated to the block length. It must always be verified that the corresponding torsional stress of the bar does not exceed the permissible yield value on the settled spring.

4.2.12 Static load "P_1"

It's the load corresponding to the weight of the vehicle stopped at full load. For cars (light vehicles) and cars derivatives it's the one corresponding to the standard weight.

4.2.13 Dynamic load "P_2"

It's the load corresponding to the length L_2 of the spring. This load is used to control the maximum expected working load.

4.2.14 Flexibility of the spring "C"

It's the ratio between the elastic variation displacement Δf of the spring and the variation ΔP of the corresponding loads. It is expressed in mm/N.

$$C = \frac{\Delta f}{\Delta P}$$

Flexibility

Loads for flexibility control must be detected 25 mm above and 25 mm below the spring length L_1. Therefore the elastic displacement of the spring, expressed in mm, is

$$\Delta f = (L_1 + 25) - (L_1 - 25)$$

The contact of the end coils with the adjacent ones must take place outside the flexibility control field.

4.2.15 Tolerances

4.2.15.1 Tolerance on the diameter "d" of the round bar For finished steel round bars on center-less grinders, the tolerance on diameter "d" is:

- ±0.05 mm for d < 25 mm,
- ±0,08 mm for d ≥ 25 and d < 35 mm
- ±0,10 mm for d ≥ 35 and d < 50 mm.

4.2.15.2 Tolerance on the internal diameter "D_i" of the spring

Tolerance on D_i = ±1%

4.2.15.3 Unwinded profile of the end of the spring
(for springs with type ends 1 and 2)

- Tolerance on the inclination "a" of the coil end tolerance on a = ±3 mm

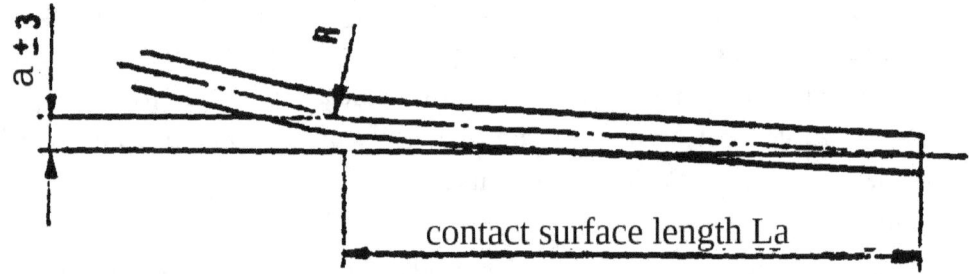

Figure 2-tolerance on inclination "a" of the coil end

- Flatness tolerance of the coil end

A deformation of 2% on the length of the supporting surface is allowed.

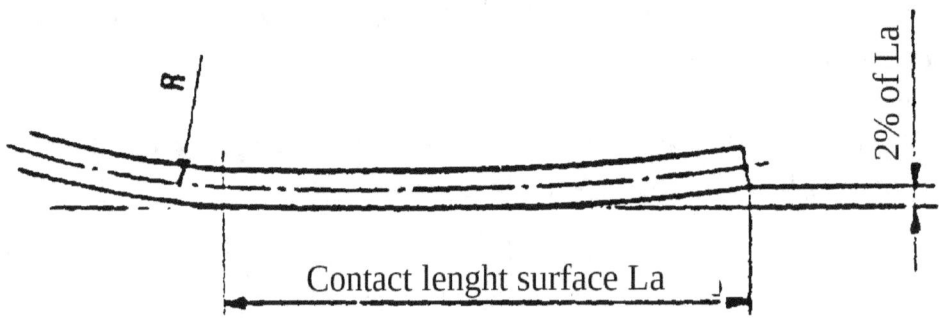

Figure 3-planarity tolerance of the coil end

4.2.15.4 Tolerance on the thickness "b" of the tapered coil end
(for springs with type 3 and 4 ends) tolerance on b = 0/-3 mm

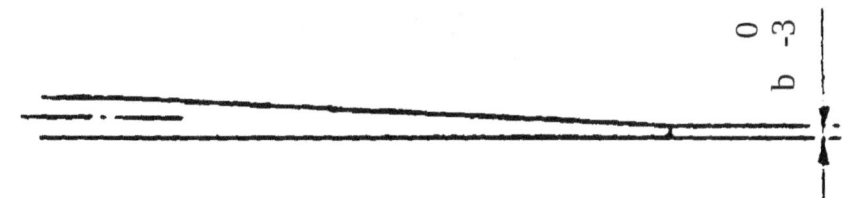

Figure 4-Tolerance on the thickness "b" of the tapered coil end

4.2.15.5 Tolerance on the tangential projection "s" of the spring end
(for spring type 1) tolerance on s = 0/-3 mm

4.2.15.6 Tolerance on the total number of coils It
For spring type 1 and 2, the tolerance is given on the angle α relative to the mutual position of the end of the coils (see example of graphical representation - Figure 7). In general its value is chosen in the field given by the following formula:

$$\Delta\alpha = 1 \, (1,8 + 2,2) It$$

Values smaller than $\Delta\alpha$ and in any case not less than $\Delta\alpha = +1,2 \, I_t$, can be chosen in case of particular needs for installation of the spring, bearing in mind that this requirement may need the cutting off of the end of the coils. For type 3 and 4 springs, the tolerance is given by the following table:

Total coils number It	Tolerance
4	±1/4 coil
4 ÷ 8	±1/2 coil
8 ÷ 15	±3/4 coil
15 ÷ 25	±1 coil

4.2.15.7 Tolerance on form error

(for springs with type 1 and 2 ends) The spring is mounted between two cups and loaded until it reaches the block length L_b. From the start of the test and throughout the excursion, the spring body must not exceed the shape cylinder having the diameter D_{max}.

$$D_{max} = p \, D_e$$
$$D_e = D_{inom} + 2 \, d_{nom}$$

Where:
D_{inom} = nominal value of D_i
d_{nom} = nominal value of d

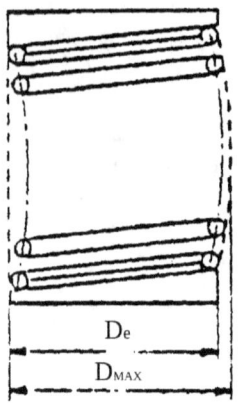

Figure 5-tolerance on form error

The value of the coefficient p is defined according to the spring slenderness ratio D_e/L_o, according to the following table:

D_e/L_o	p
≥ 0,2 and < 0,3	1,055
≥ 0,3 and < 0,5	1,030
≥ 0,5 and < 1	1,020
≥ 1	1,010

Figure 6-value of the coefficients p according to the slenderness

4.2.15.8 Tolerance on parallelism

(for springs with type 3 and 4 ends) The maximum permissible error on parallelism is normally 3°. It's controlled on the free spring. In case of particular need, the tolerance must be prescribed by design.

4.2.15.9 Tolerance on orthogonality

(for springs with type 3 and 4 ends) The maximum permissible error on the orthogonality is normally 3°. It's controlled on the free spring. In case of particular need, the tolerance must be prescribed by design.

4.2.15.10 Tolerance on static load P_1

The flexibility of the spring, the static load and the suspension trim variations, that the tolerance can cause, contribute in determining the tolerance on the static load. For this reason, the tolerance is established, case by case, or according to the flexibility (formula A) or according to the static load (formula B).

$$\Delta P_1 = \pm \frac{6 \div 8}{C_{nom}}$$

Formula A

$$\Delta P_1 = (0.04 \div 0.06) P_1$$

Formula B

The values of ΔP₁ are calculated with both formulas A and B;

- The formula that gives the value of smaller ΔP₁ is adopted;
- Among the values included within the limits of the chosen formula, the Technical Department prescribes that tolerance on the P₁ that fully satisfies the requirements of the suspension in question.

4.2.15.11 Tolerance on flexibility "C"

For springs constructed with finished round bars on center-less grinders, the tolerance is of ±3%.

4.3 example of spring suspension designation

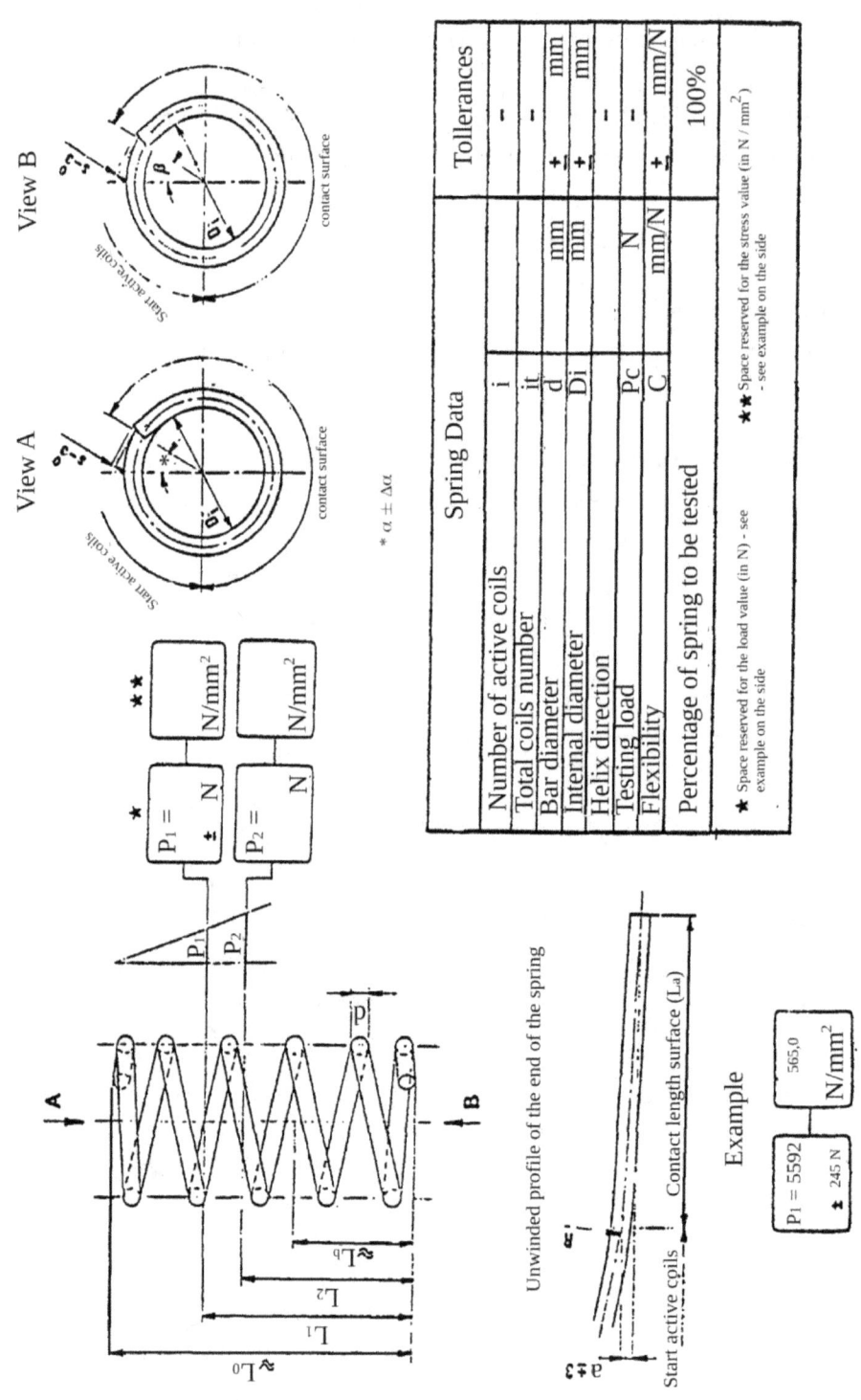

Figure 7-example of spring suspension designation

5.0 CALCULATION NOMOGRAM

5.1 Purpose

We want to provide the designer with a nomogram for the graphical calculation, as a first approximation, of cylindrical compression helix spring with steel wire circular cross-section.

5.2 Applicability

The nomogram is valid for the calculation of cylindrical compression helix spring with a circular cross-section steel wire, subjected to compression or axial traction.

5.3 Characteristics identified by the nomogram

- F' = displacement (in mm) of a spring coil under load F (in Newton);
- i_t = total number of spring coils;
- i = total number of active spring coils;
- $f = f' \cdot i_t$ (in mm): total spring displacement.

The nomogram is drawn for materials with a torsion elastic modulus G = 81400 N/mm^2

For materials having a torsion elastic modulus G' different from G, the value of the total displacement of the spring will be equal to the one calculated for G = 81400 N/mm^2 multiplied by the G/G' ratio:

- $f^* = f\, G/G' = f\, 81400/G'$ (mm). total spring displacement with torsion spring modulus G'.
- D = Spring winding average diameter (in mm).
- τ_i = Torsional stress (in N/mm^2) for springs under constant load or quasi-static conditions.
- τ_k = Torsional stress (in N/mm^2) for springs under variable load. The nomogram provides a result applicable to light vehicles.
- d = Nominal diameter of the wire (in mm) with circular cross-section.
- F = Axial load to which the spring must be subjected (in N).
- c = Winding ratio (D/d): ratio between the average winding diameter and the nominal diameter of the spring wire.
- h ed I = Construction lines of the nomogram.

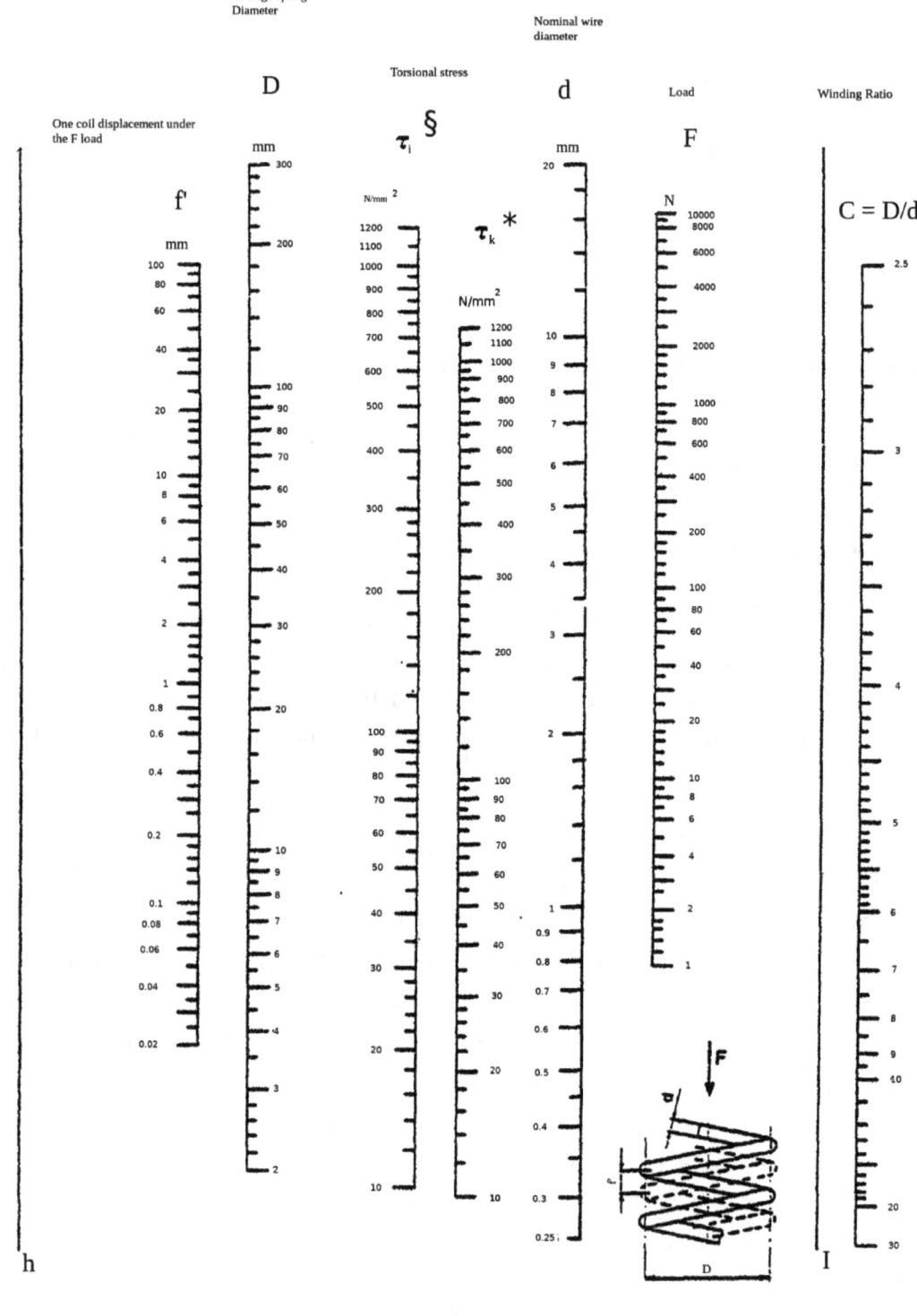

§ for springs under constant load or working in quasi-static conditions
* for springs under variable load
$G = 81400 \ N/mm^2$: Torsion elastic module

Figure 8-cylindrical compression helix spring calculation nomogram

5.4 Calculation Example

Data for a cylindrical compression helix spring:

- Average winding diameter D = 30 mm
- Nominal wire diameter d = 4 mm
- Axial load F = 100 N
- winding ratio c = 7.5
- torsion elastic module G = 81400 N/mm^2

Determine by the nomogram:

- The torsional tension (for static load) τ_i
- The torsional tension (for variable load) τ_k
- The displacement of a one coil f'

Solution:

1. Locate the corresponding data points on the respective scales (A-B-C-D).
2. Draw the line passing through A-B until you meet the construction lines of the nomogram h and I, identifying the meeting points respectively E and F
3. Draw the line passing through F-C until it meets the scale of τ_i. The meeting point H identifies on the scale the value of the torsion stress (for constant load) of the spring τ_i = 120 N/mm^2; The first question has thus been solved.
4. Draw the line H-D and locate the intersection point with the scale of τ_k. This point of intersection M identifies on the scale the value of the torsional stress (for variable load) of the spring τ_k = 140 N/mm^2; the second question was thus solved.
5. Draw the straight line E-H and find the intersection point with the scale of f'. This point of intersection N identifies on the scale the value of the displacement for each coil of the spring f'=1,1 mm; The third question was thus resolved.
6. Multiplying the value of f' by the total active spring coils number, we have the total displacement of the spring f = f' i.

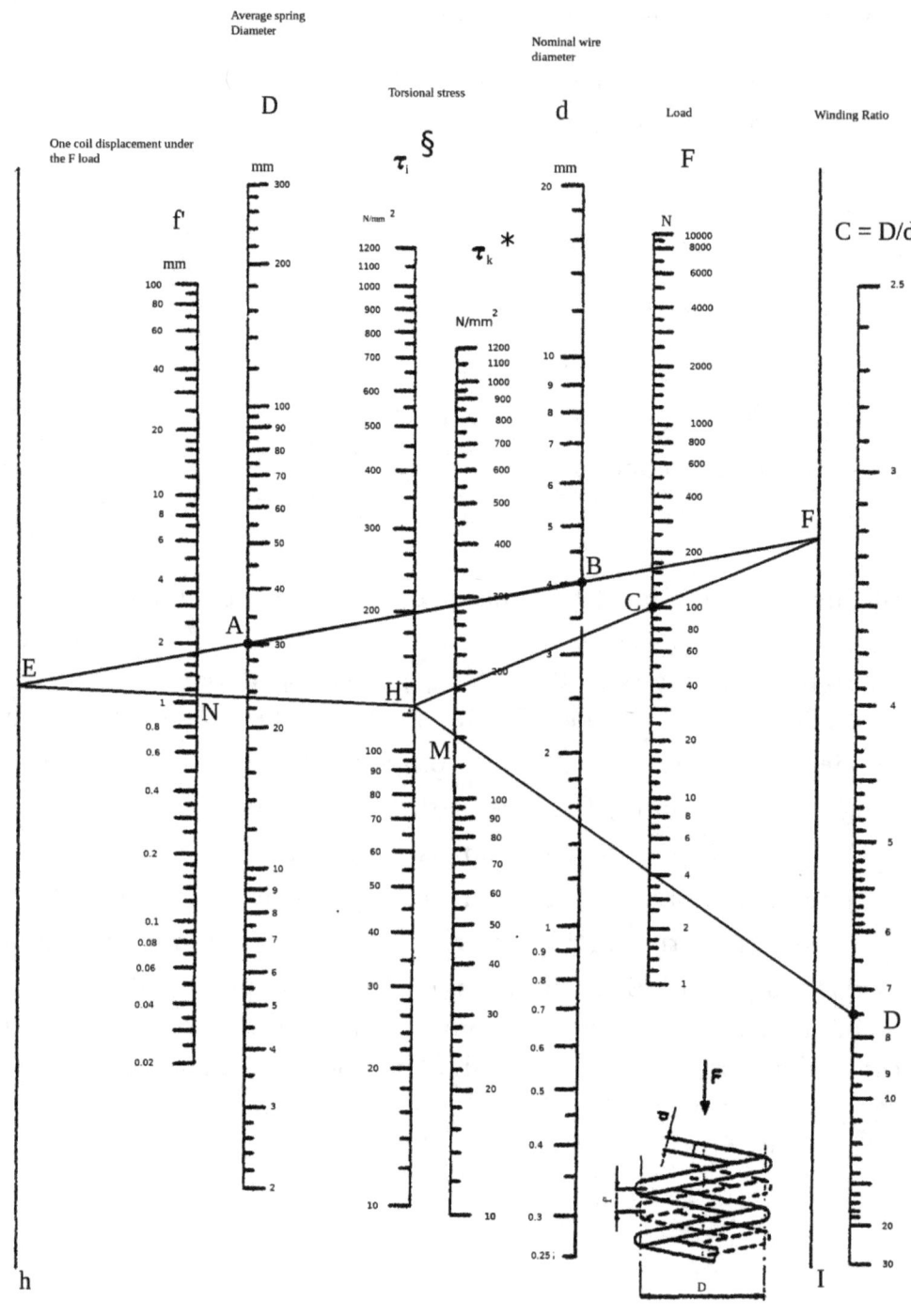

§ for springs under constant load or working in quasi-static conditions
* for springs under variable load
$G = 81400$ N/mm^2 : Torsion elastic module

Figure 9-example using the nomogram

www.ingramcontent.com/pod-product-compliance
Lightning Source LLC
Chambersburg PA
CBHW062345220526
45469CB00008B/2851